Simple Machines

Copyright © by Harcourt, Inc.

All rights reserved. No part of this publication may be reproduced or transmitted in any form or by any means, electronic or mechanical, including photocopy, recording, or any information storage and retrieval system, without permission in writing from the publisher.

Requests for permission to make copies of any part of the work should be addressed to School Permissions and Copyrights, Harcourt, Inc., 6277 Sea Harbor Drive, Orlando, FL 32887-6777. Fax: 407-345-2418.

HARCOURT and the Harcourt Logo are registered trademarks of Harcourt, Inc., registered in the United States of America and/or other jurisdictions.

Printed in Mexico

ISBN 978-0-15-362052-2
ISBN 0-15-362052-8

5 6 7 8 9 10   0908   16 15 14 13 12 11

4500332548

Visit *The Learning Site!*
www.harcourtschool.com

Lesson 1

How Do Simple Machines Help People Do Work?

VOCABULARY
work
simple machine
lever
fulcrum

Work is the use of force to move an object from one place to another.

A **simple machine** has few or no moving parts. This wheelbarrow is a simple machine.

2

A **lever** is a bar that pivots, or turns, on a point that does not move. This fixed point is called the **fulcrum**. A hockey stick is one kind of lever.

READING FOCUS SKILL
MAIN IDEA AND DETAILS

A main idea is what the text is mostly about. Details tell more about the main idea.

Look for details about simple machines and levers.

Work and Simple Machines

We use the word *work* every day. But in science, this word has a special meaning. **Work** is the use of force to move an object from one place to another. The object must move in the same direction as the force used to move it. Lifting a box is work. But carrying a box is not work. That's because the box moves sideways with you as you carry it.

The girl does work to lift the dog.

A pry bar is used to lift nails. ▶

We use machines to help us do work. A **simple machine** is a machine with few or no moving parts. You apply just one force to make a simple machine work.

The pry bar is a simple machine. You slide one end of the pry bar under a board. Then you press down on the other end. The pry bar changes the direction of this force and lifts the board. This simple machine makes work easier to do.

 How does a simple machine like a wheelbarrow change the way work is done?

◀ The boy uses a wheelbarrow to lift leaves.

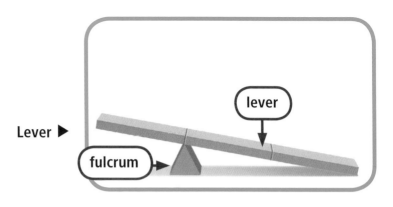

Lever ▶

Levers

A **lever** is a bar that pivots, or turns, on a point that does not move. This fixed point is called the **fulcrum**. A seesaw is one kind of lever. The board of the seesaw is the bar. The point in the middle is the fulcrum. When you push down on one end, the other end goes up. A wheelbarrow and a pry bar are two other kinds of levers.

 Tell why this hand truck is a lever.

◀ Find the lever and fulcrum in this hand truck.

◀ Hedge clipper

Levers with Other Simple Machines

Different simple machines may be used together. Look at the machine in the picture. Its handles are levers. Its blades are another simple machine called a wedge. These levers and wedges work together to make a machine that cuts bushes.

 Describe another machine found in classrooms that works like a hedge clipper.

Review

Complete this main idea statement.

1. A _____ has few or no moving parts and needs just one force to make it work.

Complete these detail statements.

2. A simple machine makes work _____ to do.

3. A lever is a bar that turns on a fixed point called a _____ .

Lesson 2

VOCABULARY
pulley
wheel-and-axle

How Do a Pulley and a Wheel-and-Axle Help People Do Work?

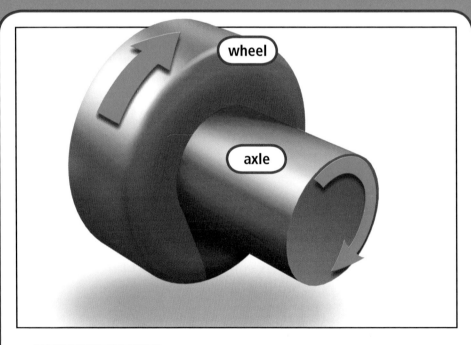

A **wheel-and-axle** is a simple machine that has a wheel and an axle that turn together. A doorknob is an example of a wheel-and-axle.

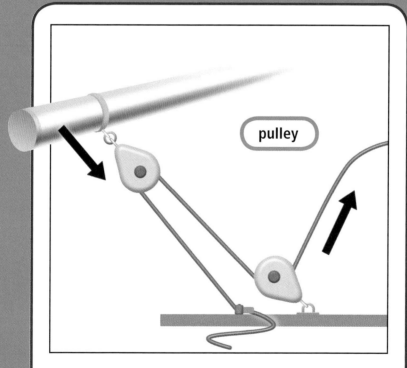

A **pulley** is a wheel with a line around it. This simple machine makes it easier to lift things.

READING FOCUS SKILL
MAIN IDEA AND DETAILS

A **main idea** is what the text is mostly about.
Details tell more about the **main idea**.

Look for **details** about pulleys and wheel-and-axles.

Pulleys

A **pulley** is a wheel with a line wrapped around it. The line may be a cord, a rope, or a chain. Small pulleys are used to raise shades. Large pulleys are used to lift machines and other heavy objects.

Like all simple machines, a pulley changes the way work is done. It changes the direction of the force. If you pull down on one end of the line, the other end goes up.

 Tell how a pulley changes the way work is done.

Wheel-and-Axle

A **wheel-and-axle** is made of a wheel and an axle joined together. An axle is the bar on which the wheel turns. To be a simple machine, these parts must turn together. When you turn the wheel, the axle turns with it.

A faucet is a wheel-and-axle. When you turn the handle, the axle turns. But you use less force than if you turned just the axle. This makes work easier.

A faucet has a wheel and an axle that turn together. ▶

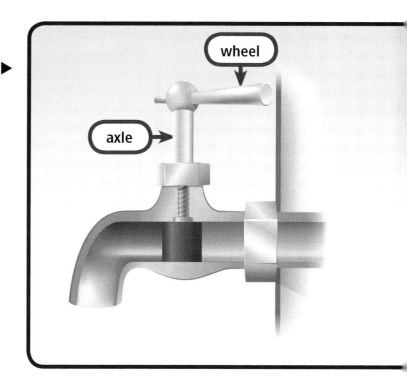

A salad spinner is another wheel-and-axle. The basket is the wheel. The crank is the axle. When you turn the axle, the basket spins. You have to move the axle just a short distance to move the outside edge of the basket a greater distance. You use more force to turn the crank, but you don't have to move it very far.

 Tell what a wheel-and-axle must do to be a simple machine.

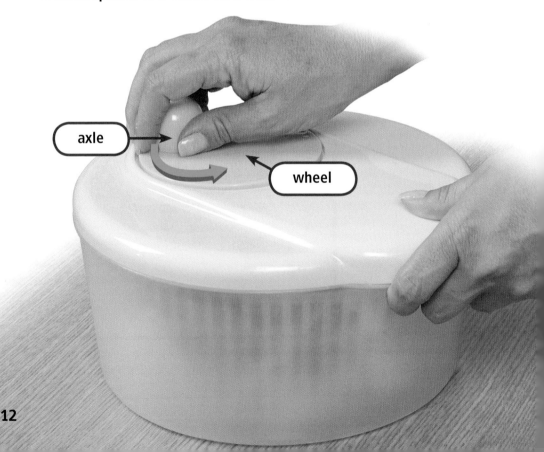

A salad spinner is a wheel-and-axle. ▼

Machines Working Together

Pulleys and wheels-and-axles can work together with other simple machines. A fishing rod and reel are one example. The fishing rod is a lever. The reel is a wheel-and-axle. When the boy turns the crank, the reel spins. This winds up the fishing line at the end of the rod.

 Is the rod and reel one simple machine? Explain.

Rod and reel ▲

Review

Complete this main idea statement.

1. A pulley and a wheel-and-axle are both _____.

Complete these detail sentences.

2. A _____ makes work easier by changing the direction you pull on a line to move something.

3. The two parts of a wheel-and-axle are joined so they always move _____.

Lesson 3

VOCABULARY
inclined plane
screw
wedge

How Do Other Simple Machines Help People Do Work?

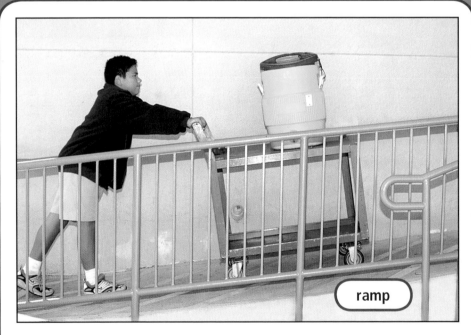

An **inclined plane** is a slanted surface, such as a ramp.

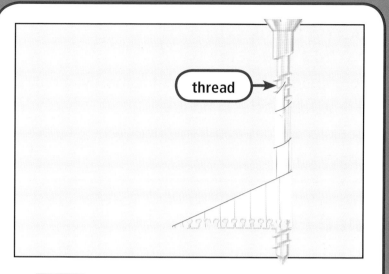

A **screw** is a post with an inclined plane wrapped around it.

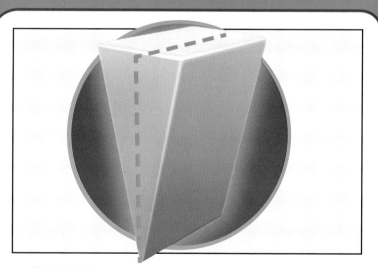

A **wedge** is two inclined planes put back-to-back. A wedge is used to split or cut things.

READING FOCUS SKILL
MAIN IDEA AND DETAILS

A **main idea** is what the text is mostly about. **Details** tell more about the **main idea**.

Look for **details** about inclined planes, screws, and wedges.

Inclined Planes

There are two ways to get this boat onto the trailer. One way is to lift it straight up. The other way is to use a boat ramp, which is an inclined plane. An **inclined plane** is a slanted surface. It is another example of a simple machine.

An inclined plane changes the way work is done. It changes the amount of force. It also changes the direction of the force. The boat travels farther, but it takes less force than lifting it straight up.

▼ A boat ramp is an inclined plane.

A hill is an inclined plane. Look at these bikers. Both ride up inclined planes to get to the top. But one bike path is much steeper than the other. The steeper path is shorter. But the rider must pedal harder. The rider on the less steep path uses less force. But that rider must travel farther. In both cases, riding up an inclined plane is less work than lifting the bike straight up.

 Tell how an inclined plane changes the way work is done.

▼ Which inclined plane is longer? Which takes more force to climb?

Screws

A **screw** is a post with threads wrapped around it. These threads are an inclined plane that curls around the post.

A drill bit is an example of a screw. By turning the bit around, you use less force than pushing the bit straight into the wood.

A nut and a bolt are also examples of screws. Their threads slide along each other. The nut and bolt keep two pieces of wood together.

Focus Skill **What is a screw?**

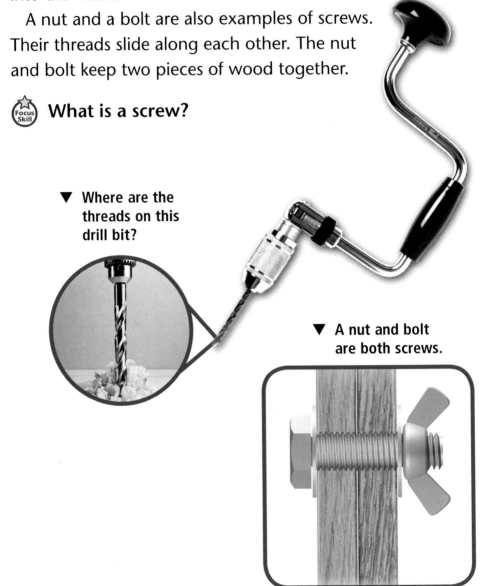

▼ Where are the threads on this drill bit?

▼ A nut and bolt are both screws.

Wedge

Like screws, wedges also use inclined planes. A **wedge** is two inclined planes put back-to-back.

You know that an inclined plane changes the direction of a force. A wedge does this, too. When you push down on a knife to cut an onion, the two parts of the onion move sideways, away from each other.

 Tell how a wedge is like an inclined plane.

◀ A knife is a wedge.

Review

Complete this main idea statement.

1. The inclined plane, screw, and wedge are all _____.

Complete these detail statements.

2. Every inclined plane is a _____ surface.

3. The threads of a screw are really an _____ wrapped around a post.

GLOSSARY

fulcrum (FUHL•kruhm) The fixed point on a lever

inclined plane (in•KLYND PLAYN) A simple machine that is a slanted surface

lever (LEV•er) A simple machine made of a bar that pivots on a fixed point

pulley (PUHL•ee) A simple machine made of a wheel with a line around it

screw (SKROO) A simple machine made of a post with an inclined plane wrapped around it

simple machine (SIM•puhl muh•SHEEN) A machine with few or no moving parts

wedge (WEJ) A simple machine made of two inclined planes placed back-to-back

wheel-and-axle (weel•and•AK•suhl) A simple machine made of a wheel and an axle that turn together

work (WERK) The use of force to move an object from one place to another